Shanghai's International Zone

A Nexus of War, Intelligence, and Survival

Pacific Atrocities Education

Shanghai's International Zone

A Nexus of War, Intelligence, and Survival

M. Stock

Shanghai's International Zone

A Nexus of War, Intelligence, and Survival

Written by
M. Stock

Editor
Samantha Lee

Published by Pacific Atrocities Education

Printed in the United States of America.

Paperback ISBN: 978-1-947766-65-5
E-book ISBN: 978-1-947766-64-8

Table of Contents

Timeline 1

Introduction 5

Chapter 1. A Difficult Battleground 19

Chapter 2. Alphabet Soup 27

Chapter 3. Civilian Intelligence 39

Conclusion 49

Works Cited 57

Timeline

A Very Long Battle

To understand the Battle of Shanghai, a months-long battle that many consider the first battle of the Pacific Theater, one must understand the events and key actors who were instrumental in the control and destruction of Shanghai during World War II. The battle lasted from the 13th of August to the 26th of November 1937 and is counted as one of the longest drawn-out battles in modern military history. The Battle of Shanghai appears on this list along with other notable battles including the Battle of Verdun that occurred during World War I, the Siege of Leningrad that happened during the Second World War, and the Battle of Ramadi during the Iraq War.[1]

Civil War in the Wake of Revolution

In the midst of the Chinese Civil War, the Japanese prepared to strike China after having successfully con-

1. "What Was the Battle of Verdun?," Imperial War Museums, accessed August 12, 2024, https://www.iwm.org.uk/history/what-was-the-battle-of-verdun.

quered other neighboring countries. The Chinese Civil War was a continuation of the Chinese Rebellion of 1911, which led to the downfall of the Qing dynasty and the Chinese Imperial system. This rebellion was led by Nationalist Party founder Sun Yat-Sen, who is often called "the father of modern China", leading to the first provisional presidency of the Republic of China. Sun Yat-Sen had both Mao Zedong and Chiang Kai-Shek within his party until his untimely death. His untimely demise from cancer in Beijing on March 12th, 1925 fractured the Nationalist Party and led to the outbreak of the Civil War between the former party members. The Chinese Civil War began in the early-mid 20th century with scattered fighting throughout China between the Communists (led by Mao Zedong) and the Nationalists (led by Chiang Kai-Shek). The conflict lasted until 1949, after which it ended with a Communist victory and the establishment of the People's Republic of China.

The Chinese Civil War was a major reason that the Japanese were able to set up a foothold on the Asian mainland as the infighting between the Nationalists and the Communists led to heavy Chinese casualties, making defending cities such as Shanghai and Nanjing during the Battle of Shanghai and the Nanjing Massacre difficult.[2] The infighting amongst the Chinese proved to be the perfect distraction for the Japanese to infiltrate China relatively undeterred until the Japanese

2. "The Chinese Revolution of 1911," Office of the Historian, U.S. Department of State, accessed April 27, 2024, https://history.state.gov/milestones/1899-1913/chinese-rev.

had captured significant amounts of territory in China. The areas captured were primarily located in northern China including Manchuria, which was used to set up the non-recognized or so-called "puppet" state of Manchukuo. This captured territory is where plans for the Battle of Shanghai were drawn up with assistance and input from Imperial Japanese Army members stationed elsewhere.

The Battle and Its Aftermath

The preparation for the battle dates back to 1932 with the first Battle of Shanghai, a battle between the Chinese and Japanese forces that lasted from January 28th to March 3rd, 1932. This battle broke out because of rising tensions between the Chinese and Japanese communities in Shanghai. The Empire of Japan sent ground and naval reinforcements to protect their countrymen from anti-Japanese sentiment in Shanghai during this time, sparking conflict.[3] During the same year of 1932, the Empire of Japan established a puppet state in Manchuria where the Japanese Army began planning for their second attempt to take Shanghai. A puppet state is defined as a state or government that seems independent but is actually controlled by another country and often has no sovereign control over the territory.

3. "The First Battle of Shanghai 1932," Pacific Atrocities Education, accessed August 16, 2024, https://www.pacificatrocities.org/the-first-battle-of-shanghai-1932.html.

Their plans were put into action following the Marco Polo Bridge Incident on July 7th, 1937 with the attack on Shanghai beginning just over a month later.

The Battle of Shanghai finally ended in late November 1937 with a Japanese victory. The Japanese assumed control of the city except for the foreign-ruled International Settlement, which did not fall to the Imperial Japanese until 1941. The International Zone remained in the hands of the British and their municipal council, as it had been since the formation of the International Zone in the early 19th century, until its takeover by the Imperial Japanese Army, signaling the total fall of Shanghai to the Japanese.[4] After the Battle of Shanghai, the Imperial Japanese Army assumed control over the city. At the same time, a portion of their forces moved south to the city of Nanjing and other cities in central and southern China. Shanghai remained in the hands of the Empire of Japan, including the International Settlement, from 1941 onwards until the end of the Asia-Pacific War in 1945.

4. Yang, Liping, "Researching the History of Shanghai Between the 1830s and 1950s," *The Gale Review*, accessed June 19, 2024, https://review.gale.com/2023/12/05/researching-the-history-of-shanghai-between-the-1830s-and-1950s/.

Introduction

The Founding of Manchukuo and the Downfall of the Qing Dynasty

Pertaining to those involved in the Battle of Shanghai, one must step back and look at the history of the region prior to the outbreak of the battle itself. Imperial Japan had been on the rise since the conclusion of the Russo-Japanese War, which ended with a Japanese victory in 1905. After, the Empire of Japan set its sights on China and the rest of Asia prior to the official beginning of the Second World War, also known as the Asia-Pacific War, the Pacific Theatre, or the War in The Pacific. For the purposes of this paper, these terms will be used interchangeably, with the Second World War primarily being used to refer to the broader world war as well as to refer to the European Theatre while the other terms will refer specifically to the Asia-Pacific region and conflict. With the intent to colonize the remainder of East Asia and Southeast Asia, the Imperial Japanese Army set about occupying various countries, including plans for setting up a foothold in China beginning in 1927. This goal came to fruition after the Mukden Incident.[5] This was a false flag planting that occurred on September 18th of 1931 and served as a justification for the Imperial Japanese invasion of Manchuria where they established the puppet state of Manchukuo and planned for their attacks on central and southern China.[6]

The independent state of Manchukuo had been known formally as "The State of Manchuria" from its establishment in 1932 until being renamed in 1934. From 1934 until 1945, it was called "The Empire of Great Manchuria" and was ruled by Henry Puyi.[7] A puppet of the Imperial Japanese, Puyi was the last Qing emperor prior to the downfall of the 2,000 year Chinese Imperial system with the Chinese Revolution of 1911.[8] Puyi had transitioned from being the Xuantong emperor of Imperial China to serving as a puppet emperor for Japan due to his Manchurian heritage after his abdication from the Imperial throne in 1912.[9] The abdication and downfall of Puyi was the end of a 2,000 year imperial rule in China with the rise of the Chinese Revolution. While memoirs and other stories exist from and about the life of Puyi, there is still much that is not

5. "Establishment of Manchukuo," Pacific Atrocities Education, accessed August 12, 2024, https://www.pacificatrocities.org/establishing-manchukuo.html.
6. GHQ US Army Forces, Pacific Military Intelligence Section, "Saionji-Harada Memoirs Part 2: Mukden to Shanghai," *Pacific Asia War Archive*, accessed July 16, 2024, https://pacificatrocitiesedu.reclaim.hosting/admin/items/show/821.
7. "Establishment of Manchukuo," ibid.
8. "The Chinese Revolution of 1911," ibid.
9. *L'ultimo imperatore*, Roma: Gruppo editoriale L'Espresso, 2003.

known regarding the full extent of his participation in Manchukuo during his time as a puppet emperor. It is known that he was manipulated at least to an extent by the Imperial Japanese and that Manchukuo was under the strict control of the Japanese, who used it as a base for further expansion into Asia. However, how much Puyi himself was aware of regarding the more secretive operations of the Japanese intelligence operatives and military police (known as the Kempeitai) and other imperial Japanese forces is less clear. We do know that after his abdication from the Chinese Imperial throne, he resided in Tianjin, a Japanese colony. From there he became known as the "Kangde of Manchukuo" in the mid-1930s, a rule that lasted until 1945 with the downfall of Manchukuo, the surrender of the Empire of Japan, and the capture of Puyi by Soviet Red Army forces. Puyi was later released from captivity and died a common citizen of China in Beijing in the late 1960s.

Before the downfall and dissolution of the Qing dynasty, the Manchu, one of the indigenous Chinese ethnic minorities that descended from the Jurchen people and citizens of Manchuria, had served in the Chinese Imperial system for hundreds of years. The Manchu wielded significant influence both in and out of the Imperial court, making Manchuria a place of interest to the Japanese considering the status of Puyi at the time.[10] Its railway system, proximity to major

10. Rawski, Evelyn S., "Reenvisioning the Qing: The Significance of the Qing Period in Chinese History," *Journal of Asian Studies* 55(4), doi:10.2307/2646525.

cities such as Shanghai and Peking (Beijing), as well as its abundance of natural resources including coal and natural gas motivated the Imperial Japanese to swiftly set about laying claim to the region and installing Puyi as emperor of what became the first of multiple puppet states throughout Asia. Manchukuo would be used as a blueprint of sorts by Japan when creating puppet states and illegitimate governments in conquered regions particularly in terms of industrial power due to Manchuria's wealth of natural resources and railway system.

Puyi, c. 1915-1920, the last Qing Emperor and puppet emperor of Manchukuo.[11]

11. "Puyi." *Encyclopedia Britannica*, April 3, 2024. https://www.britannica.com/biography/Puyi.

Crimes in Manchukuo and the Imperial Japanese Empire

After their capture of the northern region, the Japanese faced both the Nationalist and Communist armies during the Asia-Pacific War because the two Chinese factions had united to face the oncoming Japanese. At the same time, the Empire of Japan set about continuing its efforts to build an empire spanning the Pacific. The full-scale invasion of China did not begin until 1937; however, experiments involving chemical and biological warfare—some of which were deployed during the Battle of Shanghai—as well as Japanese troop mobilization and the subjugation of Manchukuo's population had already started as early as 1933. The Kamo Detachment, also known as Unit 731 "The Forgotten Asian Auschwitz", operated in Manchukuo from 1936 until 1945 and is responsible for at least 3,000 casualties with estimates ranging from 3,000 to 300,000 due to the extensive experiments involving biological warfare.[12] There were few to no survivors of Unit 731. Manchukuo was operated by officials from the Empire of Japan and ruled symbolically by the former Qing emperor Puyi, though he had little real power.

Prior to the Japanese invasion and occupation, Manchuria was ethnically diverse with populations of Han Chinese, Manchurians, and other non-Han Chinese ethnic minorities, Koreans, Japanese, Russians,

12. Pua, Derek, Dybbro, Danielle, and Rogers Alistair, *Unit 731: The Forgotten Auschwitz* San Francisco, CA: Pacific Atrocities Education, 2018.

and others. Manchukuo was an unrecognized independent Japanese state in Manchuria used as a staging ground for the invasion of mainland Asia, which was such an integral component in Japanese military strategy that without it, they potentially, if not likely, would have failed to capture China. The ethnic minorities of Manchukuo were used by the Imperial Japanese as slave labor and for human experimentation. The brutality of the experimentation and subjugation was on par with that of the German concentration camps such as Auschwitz. Unlike the infamy of the German concentration camps and their crimes including the murder of six million Jews and six million others including gay men, disabled people, and the nomadic Romani people, the atrocities that the imperial Japanese army committed in China and the rest of Asia during the Second World War are often overlooked.[13]

While many know of the crimes committed in Europe, few know of the horrors that the colonial states of the Empire of Japan inflicted on territories under Japanese occupation. For example, in Japanese-occupied Korea, there were frequent examples of sexual crimes and slavery. In the Philippines, there was the Bataan Death March where Filipino and American soldiers who became Japanese prisoners of war were forcibly marched to Camp O'Donnell with the weak slaughtered along the way. Japanese colonial states included Malaysia, Singapore, Thailand, Cambodia, Vietnam, and Laos, in addi-

13. "What Was the Holocaust?" Yad Vashem, accessed August 14, 2024, https://www.yadvashem.org/holocaust/about.html.

tion to mainland China.[14] Other places in the Pacific region, while not colonized, were subject to attempted bombardment such as the Midway Atoll, a ring-shaped chain of islands formed of coral with a lagoon in the center, and portions of northern Australia—though these places could not be successfully captured. The Allied military base at Darwin in northern Australia was attacked on the 19th of February 1942 by Imperial Japanese Vice Admiral Nobutake Kondo only months after the Japanese attack on Pearl Harbor in Hawaii. The attack on Darwin was an attempt to disrupt Allied transmissions, intelligence activities, and shipping routes, though it was not captured for the purposes of colonization—unlike many countries and territories in the Pacific.[15]

In all cases, bombarded and/or occupied territories endured varying forms of harsh rule while the Japanese systematically took control of much of East and South-East Asia.[16] They ruled with an iron fist while they continued planning to conquer elsewhere. The colonization tactics of the Empire of Japan in the Pacific The-

14. Reference Division, "Developments in the Economic Structure of the Co-Prosperity Sphere During the Past Year Part II—The Outer Zone," *Pacific Asia War Archive*, accessed June 12, 2024, https://pacificatrocitiesedu.reclaim.hosting/admin/items/show/714.
15. "1942: February 19: Japanese Raid at Darwin, Australia," accessed August 18, 2024, https://www.history.navy.mil/content/history/museums/nmusn/explore/photography/wwii/wwii-pacific/us-entry-into-wwii-japanese-offensive/1942-January-15-abda-japanese-octopus/1942-february-19-japanese-raid-darwin-australia.html.
16. Reference Division, "Developments in the Economic Structure of the Co-Prosperity Sphere During the Past Year Part II—The Outer Zone," ibid.

ater are vast and complicated. This work will focus on the impact of military intelligence on tactics used by the Imperial Japanese as well as the strategy and tactics used by their enemies, The Allied Forces: the French, British, Americans, Russians, and Chinese.[17] Only the Chinese forces were directly engaged in combat with the Japanese during the Battle of Shanghai. However, the other Allied Forces maintained intelligence operations and contributed military intelligence in Shanghai. With a solid foothold in Manchukuo in the northeast of China, the Imperial Japanese Army was able to organize and plan an attack on the central and southern regions—beginning with the international port city of Shanghai.

Military and National Impact

Due to the manipulation of Puyi and a strong military presence, the Empire of Japan was a formidable military power that frequently engaged Chinese forces from their stronghold in Manchukuo in the northeast of China, though Manchukuo was weak from a political and nationalistic standpoint. The puppet state additionally was unrecognized by the League of Nations and subject to scrutiny on an international level despite its military prowess. The Empire of Japan, despite the weakness of

17. "US Marines Intelligence Report 5 December, 1935," accessed July 4, 2024. https://www.pacificatrocitiesedu.reclaim.hosting/items/show/221. "Central China Area Operations Record, 1937-1941", Office of the Chief of Military History Department of the Army, accessed July 4, 2024, http://ibiblio.org/hyperwar/Japan/Monos/pdfs/JM-179/JM-179.pdf.

Manchukuo politically, utilized the unrecognized state in their plan to bring Asia under their rule and the Battle of Shanghai is often thought of as the beginning of the Second World War, both in general and within the Pacific Theater. Japanese forces coming out of Manchukuo began planning for the invasion of the rest of China. A small group of soldiers amassed near the Marco Polo Bridge on July 7th 1937, which was close to a major Japanese railway site, and demanded entry into a village.[18] They were faced with resistance from the Chinese soldiers, which resulted in a shot being fired and fighting breaking out. This incident is considered an important precursor to the full-scale invasion of China after the founding of Manchukuo and the beginning of the intense Japanese involvement in China that continued until 1945 with the surrender of Japan and the end of the Asia-Pacific War. After the Marco Polo Bridge Incident, Japan mobilized their troops and attacked Shanghai in August 1937. This was the second attack on Shanghai since 1932 when the initial Japanese attack on Shanghai occurred.

The Battle of Shanghai is thought to be the true start of the Second World War by many historians. The use of chemical weapons by both Chinese and Japanese forces during the battle combined with the devastating impact of urban warfare was an important factor in the transition to all-out war with tactics heavily utilizing previously unthought of modern weapons of war. These

18. "Japan, China, the United States and the Road to Pearl Harbor, 1937-41," U.S. Department of State, accessed August 16, 2024, https://history.state.gov/milestones/1937-1945/pearl-harbor.

new weapons included chemical and biological weapons developed in Manchukuo.

The Battle of Shanghai lasted from August to November 1937, with an eventual Japanese victory due to its air and naval power coupled with the sense of its martial (military) superiority stemming from the ideas of Bushido. Since the early feudal period, Japanese warriors, specifically the samurai, had followed the code of Bushido or "the way of the warrior". This code was formalized during the Edo period in 1603 and the ideas continue to be found in Japanese culture long after the demise of the samurai with a surge in popularity during the Meiji Restoration from 1868 to 1889. With the outbreak of the Asia-Pacific War, the Japanese Army drew on the code of Bushido, in particular the pillars of courage, honor, and loyalty.[19] These values were embedded in the minds of Japanese soldiers as they fought to capture Shanghai for the glory of the Empire of Japan and to bring honor to themselves and their country. Trench warfare between the Chinese and Japanese forces occurred on the outskirts of Shanghai during the battle with the Chinese forces fleeing. A majority of the fighting, however, occurred within the city center in the form of urban warfare.[20]

---

19. Clark, Tim, "The Bushido Code: The Eight Virtues of the Samurai," accessed August 16, 2024, https://china.usc.edu/sites/default/files/forums/Samurai%20and%20the%20Bushido%20Code.pdf.
20. Paulose, James. "Three Months of Bloodshed: Strategy and Combat During The Battle of Shanghai." *Report: West Point Undergraduate Historical Review* 3, Vol. 2 (2013): 9-22, https://athena.westpoint.edu/items/ca1491c1-b912-49e7-9e38-8ec9867d13c4.

The Battle of Shanghai was a key point in Japan's expansion into China and the outbreak of the Second World War.[21] While the Imperial Japanese Army had been stationed in China since 1932, establishing the puppet state of Manchukuo in northeastern Manchuria, the Battle of Shanghai marked the first major engagement of the Asia-Pacific War. Its significance was amplified by Shanghai's large foreign population, which Chiang Kai-Shek strategically leveraged on the global stage, thrusting China—and particularly Shanghai—into the international spotlight with the outbreak of war. Chiang Kai-Shek used Shanghai as a stronghold similar to how the Imperial Japanese used Manchukuo; both armies shipped in weaponry from Germany to use against the other. Chiang Kai-Shek used Shanghai as a strategic stronghold to resist Japanese expansion, while the Imperial Japanese Army leveraged Manchukuo as an occupied territory to consolidate power in northeast China. Both Chinese and Japanese forces deployed substantial military assets, with Chiang's army utilizing weaponry and training sourced from Germany. The Marco Polo Bridge Incident marked a critical event leading up to the Battle of Shanghai, as it ignited full-scale hostilities between China and Japan shortly before the outbreak of the battle. Following the Japanese victory at the Marco Polo Bridge, the invasion of Shanghai allowed the Imperial Japanese Army to push further into central China. Shanghai proved to be a tactically challenging battle-

21. "Marco Polo Bridge Incident," *Encyclopedia Britannica*, June 30, 2024, https://www.britannica.com/event/Marco-Polo-Bridge-Incident.

ground, with intense urban combat in the city center and entrenched warfare on its outskirts. Air and naval engagements occurred sporadically throughout the city, while the International Settlement served as neutral ground until Japanese forces took control in 1941. Consequently, the Battle of Shanghai extended over several months, with continuous fighting on land, in the air, and along the waterways.

It is important to note that the Imperial Japanese Army and the United Chinese Army were not the only groups involved in the city of Shanghai or the Battle of Shanghai. The fighting during the Battle of Shanghai primarily took place between Imperial Japanese forces—comprising the Japanese Army, Navy, and Air Force—and the Chinese Army, which included both the Nationalist and Communist forces, now united, along with their small Air Force. In addition to these two primary armies, other foreign countries' armies and operatives were involved as well. The British administered the International Settlement, which provided refuge to the Shanghainese citizens fleeing the intense fighting within the decimated city. Additionally, the British oversaw their own British citizens, other citizens of the International Zone, and an influx of European refugees. The French administered the French Concession adjacent to the British-run International Settlement. Lastly, there were a number of Americans in and around Shanghai prior to, as well as during, the Battle of Shanghai. Primarily, it was members of the American Marine Corps and Office of Naval Intelligence who kept extensive records on many of the different groups, armies, and important individuals in Shanghai, especially during what was referred to as "The State of Emergency" or "The Shanghai Emergency".[22]

22. "US Marines Intelligence Report 5 December, 1935," ibid. "Central China Area Operations Record, 1937-1941", ibid.

Chapter 1

A Difficult Battleground

Neutral Ground and Urban Warfare

When attempting to capture the city of Shanghai, the ability to access and maneuver through the International Zone was seen as a distinct advantage, especially due to the International Zone's status as a neutral territory. The International Zone was classified as neutral ground and in the hands of the British municipal council until its fall and takeover by the Imperial Japanese Army in 1941. The International Zone remained under the control of the Imperial Japanese Army until the end of the Second World War with the surrender of Japan after the use of atomic bombs at Hiroshima and Nagasaki on September 2nd, 1945, also known as V-J Day where Imperial Japan surrendered aboard the US Navy vessel the USS Missouri.[23] From a military tactic perspective, the Battle of Shanghai was significant due to it primarily being urban warfare. With a dense civilian population and a large area geographically occupied,

23. "'Japan Surrenders!': The National World War II Museum reminds Americans of the 65th anniversary of V-J Day and end of WWII," The National WWII Museum, New Orleans, August 10, 2010, https://www.nationalww2museum.org/media/press-releases/japan-surrenders-1.

the urban nature of Shanghai made the avoidance of civilian casualties difficult. As a result, many Shanghai civilians fled to the International Zone for safety.[24] The neutral territory took in such a large influx of fleeing civilians that many living quarters were cramped with up to 30 people per room.

Citizens of Shanghai fleeing into Shanghai International Settlement.[25]

American Marines Arrive in Shanghai

In 1935, Major William Arthur Worton, an officer of the United States Marine Corps, reported to the Office of Naval Intelligence's Far East Section. Worton worked

24. "Citizens of Shanghai Fleeing into Shanghai International Settlement—Battle of Shanghai (1937)," PICRYL—Public Domain Media Search Engine, https://jenikirbyhistory.getarchive.net/media/citizens-of-shanghai-fleeing-into-shanghai-international-settlement-battle-6b4b10.
25. Ibid.

in intelligence alongside members of the Office of Strategic Services and Marine Corps Intelligence in Asia with the suggestion that the Office of Naval Intelligence focused on the International Settlement in Shanghai and in Hong Kong.[26] Major William Worton rose to the rank of colonel during his time in the Far East throughout World War II, serving in key areas of the Pacific Theater, including Shanghai and Okinawa.[27] During his service, a decision was made to establish a network of civilian and military intelligence agents to operate out of Shanghai and Japan. This network focused on monitoring fleet movements and ground operations in the region. Despite collaboration between U.S. forces, British intelligence, and the Chinese Army, there was little sharing of highly classified information among the parties involved in Shanghai's International Zone.

The United States Marine Corps (USMC) and the U.S. Navy conducted operations along the Yangtze River throughout the war, often under difficult conditions. Following the tragic "Bloody Saturday" bombing on August 14, 1937—when Chinese Air Force planes, aiming for a Japanese light cruiser, mistakenly dropped bombs on the densely populated Bund area of the Inter-

---

26. Noble, Dennis L., "A US Naval Intelligence Mission to China in the 1930s," *Studies In Intelligence* Vol. 50 No. 2 (2006)," Central Intelligence Agency, accessed August 17, 2024, https://www.cia.gov/resources/csi/studies-in-intelligence/volume-50-no-2/a-us-naval-intelligence-mission-to-china-in-the-1930s/.
27. "The United States Marines in North China," US Marine Corps, accessed August 23, 2024, https://www.usmcu.edu/Portals/218/TheUnitedStatesMarinesInNorthChina1945-1949.pdf.

national Settlement—the U.S.[28] Navy deployed vessels to assess the damage and manage the aftermath. The bombing, which caused numerous civilian casualties, further intensified the focus on intelligence and military activities in Shanghai as the conflict deepened.[29]

Victims by bombing of Chinese bombers—Battle of Shanghai.[30]

28. "Victims of "Bloody Saturday" bombing, Ave Edward VII, Shanghai, 14 August 1937," *Historical Photographs of China*, accessed June 24, 2024, https://hpcbristol.net/visual/AL-s63.
29. "U.S. Armed Forces in China," *Dictionary of American History*, Encyclopedia.com, August 17, 2024, https://www.encyclopedia.com/history/dictionaries-thesauruses-pictures-and-press-releases/china-us-armed-forces.
30. "Victims by Bombing of Chinese Bombers, Battle of Shanghai, 1937," PICRYL—Public Domain Media Search Engine, accessed October 12, 2024. https://jenikirbyhistory.getarchive.net/media/victims-by-bombing-of-chinese-bombers-battle-of-shanghai-1937-ce0c07.

Operations of the American Naval fleet along the Yangtze River occurred throughout the war in the effort to protect Shanghai and other key international port cities.[31] The role of the 4th Regiment of the United States Marine Corps in the early days of their mission in Shanghai was to protect the lives and property of American citizens in the International Zone because the former American Concession had been absorbed into the International Settlement during its reorganization in the late 19th century.[32] Following the establishment of Manchukuo and the beginning of Japanese attacks on Shanghai, the 4th Marines were deployed to protect the perimeter of the International Settlement. They were withdrawn two weeks before the Japanese attack on Pearl Harbor and reassigned to the Philippines, where many became POWs, leaving the International Zone vulnerable to capture by Japanese forces—which occurred the day after the attack on Pearl Harbor in Hawaii on December 8th, 1941.[33]

---

31. "Yangtze River Patrol and Other US Navy Asiatic Fleet Activities in China, 1920-1942, as Described in the *Annual Reports of the Navy Department*," Naval History and Heritage Command, accessed August 17, 2024, https://www.history.navy.mil/research/library/online-reading-room/title-list-alphabetically/y/yangtze-river-patrol-and-other-us-navy-asiatic-fleet-activities-in-china.html.
32. "The China Marines | History Detectives," PBS, accessed August 17, 2024, https://www.pbs.org/opb/historydetectives/feature/the-china-marines/.
33. Ibid.

Tactical Advantages

The Shanghai International Settlement was a strategically important area due to its unique architecture, location along the Huangpu River, and diverse, multicultural population. These attributes made it a desirable and highly fortified outpost for intelligence operations by various foreign powers, particularly compared to other regions in the Pacific. The Settlement's infrastructure, especially along the Bund, featured durable, well-constructed buildings, many of which remain today. The high standard of construction and roadways established by the Shanghai Municipal Council (SMC) enhanced its attractiveness as a social and tactical hub.

The dense, multicultural population within the International Settlement also allowed intelligence agents from different nations—American, British, and Japanese, among others—to move relatively inconspicuously, blending into the transient community. The complex mix of foreign concessions, including the independent French Concession, added another layer to covert activities and strategic movement within Shanghai.

In response to increasing tensions and conflict with Japan, British authorities set up defensive road barricades and attempted to protect key areas within the International Settlement. However, these efforts ultimately fell short, as Japanese forces overtook the International Settlement in 1941, effectively ending its neutral status. The settlement's robust layout and its social significance made it a central location for intelligence and military operations during the wartime period.[34]

34. Henriot, Christian, "August 1937: War and the Death En Masse of Civilians: Virtual Shanghai," *Virtual Shanghai*, accessed August 18, 2024, https://www.virtualshanghai.net/Texts/Articles?ID=130.

Chapter 2

Alphabet Soup

The Origins of Modern Intelligence

To understand the critical role of intelligence agents and the classified information they provided during World War II, one must consider the significant evolution of American intelligence. Initially, intelligence operations in the United States were fragmented, with uncoordinated efforts across the Army, Navy, and federal government. This landscape changed dramatically with the establishment of the Office of Strategic Services (OSS) in 1942. Led by William Donovan, the OSS coordinated intelligence analysis, covert operations, and counterintelligence, setting the foundation for a unified approach that would later influence the structure of the Central Intelligence Agency (CIA) after the war. The OSS laid the groundwork for modern intelligence, with its activities extending across Europe, Asia, and North Africa. It fostered intelligence-sharing and operational support with Allied forces, though tensions and competitive dynamics sometimes persisted between branches. Following WWII, the OSS was dissolved, and much of its personnel, resources, and mission transferred to the newly formed CIA in 1947.[35] Meanwhile, the Office of Naval Intelligence (ONI), founded in 1882, continued its operations, later forming the Naval Investigative Service (NIS) in 1966, which would eventually become the Naval Criminal Investigative Service (NCIS) in the 1990s. This evolution paralleled the efforts to streamline intelligence operations, bringing the Army's Military Intelligence Corps and the Intelligence and Security Command (INSCOM) under centralized coordination, both of which remain active in domestic and international operations today. The FBI, meanwhile, maintained intelligence responsibilities in Latin America during WWII, creating a competitive environment with the OSS. This lack of interagency cohesion illustrated the need for post-war restructuring, leading to the establishment of the CIA and later the Defense Intelligence Agency (DIA), formalizing collaboration between agencies for comprehensive global intelligence efforts.[36] These different agencies operate domestically as well as internationally in both the European and Pacific Theaters.

Each of these specialized U.S. intelligence agencies, many originating in the late 19th and early 20th centuries, underwent substantial reorganization during and after WWII. The Office of Naval Intelligence (ONI), established in 1882, played a vital role in the Asia-Pa-

35. "Between World Wars," *American Military History*, Chapter 19, accessed August 17, 2024, https://www.history.army.mil/books/AMH/amh-19.htm.
36. Reynolds, Nicholas, *Need to Know: World War II and the Rise of American Intelligence*. Boston: Mariner Books, 2023.

cific theater, particularly in the Battle of Shanghai. Marine Corps Intelligence began serious operations in Shanghai by 1932, with records showing U.S. Marines were stationed and collecting intelligence during the "Shanghai Emergency," a period of heightened military tension between China and Japan.

ONI and the Marine Corps collaborated closely, setting up intercept stations in Shanghai and Beijing to monitor Japanese communications and tactics. Although the Shanghai station briefly closed in 1929, it resumed in 1935 and provided valuable intelligence throughout WWII, especially during key conflicts like the Battle of Shanghai. These intelligence records, along with British intelligence files, were largely declassified in the 1980s and 1990s, offering new insights into the U.S. role in Asia.

Beyond Shanghai, OSS operatives fueled anti-Japanese resistance across Southeast Asia, especially in Thailand, supporting local movements to undermine Japanese influence in the region. These interagency efforts across Asia helped set the foundation for post-war intelligence coordination and showcase the evolving complexity of U.S. intelligence efforts during WWII.

The declassification of U.S. Marine Corps and intelligence records began in the 1970s and continued through the 1980s and 1990s, revealing extensive intelligence activities conducted in Shanghai and other parts of Asia during the Sino-Japanese conflict and WWII. Marine Corps Intelligence was active in Shanghai as early as the early 1930s, gathering information on Japanese activities, particularly within the International Settlement. This presence continued through the Battle of Shanghai (1937), with American and British intelligence operatives both operating in the city, though their strategies differed.

British intelligence focused on securing their sector of the International Settlement, using intelligence to counter Japanese advances while avoiding direct conflict. American operatives had more freedom to maneuver, gathering intelligence in both the British-controlled International Settlement and the adjacent French Concession. Documentation about the French intelligence presence in Shanghai during this period is sparse, as their intelligence operations were relatively limited compared to those of the U.S. and British forces.[37]

Both British and American intelligence agencies documented Japanese use of chemical and biological warfare in Shanghai. Through the Office of Strategic Services (OSS), the Office of Naval Intelligence (ONI), and the British Secret Intelligence Service (SIS), extensive reports were kept on Japanese operational strategies, particularly through intercepted transmissions. These reports, along with other intelligence records, were eventually declassified, offering new insights into Allied intelligence activities in Asia during WWII.[38]

---

37. "Counter-espionage summary—Shanghai," CIA, accessed August 19, 2024, https://www.cia.gov/readingroom/docs/DOC_0000112821.pdf.
38. "15 May 1947 WAR (Chemical Corps), to: CINCFE (G-2) (Chief Chemical Officer for Fell) NR: W 98097," Pacific Asia War Archive, accessed August 17, 2024, https://pacificatrocitiesedu.reclaim.hosting.

American Intelligence

The Office of Naval Intelligence (ONI) was integral to U.S. efforts in the Pacific Theater, focusing on counter-intelligence and monitoring Japanese naval activities. Working closely with Marine Corps Intelligence, ONI analyzed Japanese fleet movements and intercepted communications to understand Japanese intentions. However, Japan employed strict radio silence, cryptographic changes, and decoy strategies, which limited the effectiveness of U.S. intelligence and contributed to a delayed response before the Pearl Harbor attack.

On December 7, 1941, Japan launched a surprise assault on Pearl Harbor, targeting the U.S. Pacific Fleet stationed at Oahu, Hawaii. The attack began just before 8 a.m. local time and involved two waves of Japanese aircraft, including torpedo bombers, dive bombers, and fighter planes. In under two hours, the attack had inflicted devastating damage: eight battleships were hit, with the USS Arizona and USS Oklahoma suffering catastrophic losses. Overall, over 2,400 Americans lost their lives, with thousands more wounded, and nearly 200 aircraft were destroyed. This sudden strike on Pearl Harbor marked a turning point, drawing the United States into World War II.

Douglas Wada, a Japanese-American ONI operative stationed in Hawaii, contributed to pre-war intelligence efforts by monitoring suspected espionage networks. While details of his specific activities are limited, intelligence personnel like Wada were instrumental in countering potential threats within the Japanese-American community. However, there is no direct evidence that Wada or other intelligence agents had prior knowledge of Japan's Pearl Harbor plans.[39]

The Office of Naval Intelligence (ONI) played a key role in the Pacific Theater during WWII, coordinating with the Office of Strategic Services (OSS) and other agencies to support U.S. intelligence operations across the Pacific and mainland Asia. Formed in 1942, the OSS centralized intelligence and covert operations, bringing together military and civilian efforts across multiple fronts. Despite this, ONI faced challenges in working with other agencies, such as the FBI and Army's Military Intelligence Division, due to interagency secrecy and jurisdictional disputes. These coordination difficulties highlighted the need for more structured interagency cooperation.[40]

After WWII, the dissolution of OSS in 1945 led to the establishment of the Central Intelligence Group, later becoming the Central Intelligence Agency (CIA) in 1947. This reorganization aimed to prevent the rivalries and fragmentation that had hindered wartime intelligence efforts. The CIA inherited much of the OSS's structure and personnel, creating a centralized intelligence agency focused on both foreign and domestic threats as the Cold War emerged. This shift toward a unified intelligence framework underscored the importance of inter-

---

39. Harmon, Mark, and Carroll, Leon, *Ghosts of Honolulu: A Japanese Spy, a Japanese American Spy Hunter, and The Untold Story of Pearl Harbor*. New York: Harper Select, 2023.
40. Ibid.

agency collaboration, as the U.S. intelligence community adapted to new geopolitical challenges.

During the Cold War, ONI continued to work closely with other intelligence bodies, including the newly established CIA, as they expanded intelligence-sharing and coordination. This collaboration was essential in meeting the growing demands of the Cold War, especially in the areas of maritime intelligence and counterintelligence.

This progression toward a unified intelligence community reflects the OSS's foundational influence and the evolving roles of agencies like ONI within the larger framework of U.S. intelligence.[41]

British Intelligence

The British Secret Intelligence Service (SIS), also known as MI6, was founded in 1909 as part of the Secret Service Bureau to address growing concerns over German imperial ambitions. Mansfield Cumming, the first director, established foundational practices, including signing as "C," a tradition still used by SIS chiefs today. Cumming also introduced the "CX" designation for intelligence reports, which remains in use.

Although "MI1(c)" was an early designation, "MI6" became a commonly used, though informal, title during WWII.[42] It reflected SIS's expanding global operations,

41. Nicholas Reynolds, *Need to Know*.
42. "SIS—Our History," accessed August 19, 2024, https://www.sis.gov.uk/our-history.html.

particularly in Germany, Russia, and China. By the late 1930s, SIS anticipated conflict and created "Section D" to oversee sabotage and covert actions against German infrastructure. SIS intensified its operations in China as WWII loomed, though closer collaboration with the American OSS largely occurred post-war, driven by Cold War priorities.[43]

During WWII, while SIS and the OSS coordinated in Europe and North Africa, their relationship deepened significantly after the war, as shared interests in countering Soviet influence emerged. The OSS's reorganization into the CIA further solidified this relationship, laying the groundwork for the more integrated intelligence efforts seen during the Cold War.[44]

Chinese Intelligence

The Chinese Nationalist Army worked closely with the Office of Strategic Services (OSS) to combat Japanese forces during WWII, particularly through SACO (Sino-American Cooperative Organization) and OSS Detachment 202. This collaboration, led by Dai Li (head of the Nationalist intelligence agency Juntong) and U.S. Navy Commander Milton Miles, focused on guerrilla warfare, intelligence gathering, and sabotage against Japanese assets. SACO's activities included training Chinese troops in covert operations and equipping them for

43. Ibid.
44. Ibid.

sabotage missions, while OSS operatives also worked alongside Nationalist Chinese forces to organize paratrooper and guerrilla units. These units targeted Japanese supply lines, destroying critical infrastructure in Japanese-occupied areas.[45] The Juntong, established in 1938 and controlled by Dai Li, also coordinated with local triads like the Green Gang in Shanghai, leveraging their networks for intelligence collection and to eliminate Chinese collaborators working with the Japanese. These operations frequently involved assassinations of key collaborators and were part of broader efforts to maintain loyalty to Chiang Kai-shek's government.[46] While the OSS's presence in China grew by 1943, internal conflicts sometimes strained SACO's effectiveness. Dai Li often prioritized Chiang's domestic control, sometimes diverting SACO's focus from anti-Japanese operations to internal security, which occasionally conflicted with OSS objectives. The Communist Chinese Army was less involved with these intelligence operations, as Chiang Kai-shek retained control over the SACO and Nationalist-led OSS collaborations. As a result, the Nationalist forces took primary roles in these operations, with OSS support largely sidelining Communist participation.[47]

This collaborative network, despite internal tensions, significantly impacted anti-Japanese activities,

---

45. "Detachment 202," OSS Office of Strategic Services, accessed August 18, 2024, https://www.soc.mil/OSS/det-202.html.
46. Fenby, Jonathan, *Chiang Kai Shek: China's Generalissimo and the Nation He Lost*. United States: Hachette Books, 2009.
47. "Secret Intelligence," OSS Office of Strategic Services, accessed August 18, 2024, https://www.soc.mil/OSS/secret-intelligence.html.

with OSS-trained Chinese forces conducting impactful sabotage and intelligence missions in Japanese-controlled regions.

Japanese Intelligence

The Kempeitai operated as the military police of the Imperial Japanese Army (IJA), while also functioning as a civilian secret police and intelligence agency across Asia. Their role was extensive, including enforcing military discipline, conducting surveillance, managing covert intelligence, and carrying out prisoner interrogations—often with brutal methods. Known for their ruthless interrogations, the Kempeitai frequently employed torture to maintain control over occupied populations, particularly in areas like China and Southeast Asia.[48] Their operational structure allowed flexibility: Kempeitai agents worked not only in large units but often in small cells or individually, which enabled covert intelligence gathering and efficient reporting to higher authorities. In places like Shanghai, the Kempeitai's headquarters, known as "Bridge House," served as a base for intelligence operations, where they targeted resistance forces through violent repression. The organization was adept at adapting to various tasks, including intercepting com-

48. Hall, Simon, "Imperial Japanese Army Intelligence in North and Central China During the Second Sino-Japanese War," *Salus Journal,* Vol. 2, Issue 2 (2014), pp. 16-30. Office of Justice Programs, accessed August 18,2024,https://www.ojp.gov/ncjrs/virtual-library/abstracts/imperial-japanese-army-intelligence-north-and-central-china-during.

munications, conducting assassinations, and gathering intelligence on potential resistance.[49] Although intercepting communications, including telegrams, was part of their responsibilities, the Kempeitai also relied heavily on forced confessions and local informants to gather information. While they occasionally coordinated with Japanese diplomatic missions or naval intelligence, the Kempeitai often operated autonomously, particularly in territories under direct Japanese control. Their combination of military, police, and intelligence functions, along with their calculated use of terror and intimidation, made them a feared presence in Japanese-occupied territories.

Modes of Operating

Each of the armies employing intelligence agents operated in their unique ways, adapting to the demands of their respective theaters and missions. Japanese intelligence, led by the Kempeitai and supported by agents from the Naval General Staff, primarily conducted covert operations in small cells or individually. Takeo Yoshikawa, for example, operated undercover as a diplomat in Hawaii, gathering intelligence on U.S. naval movements at Pearl Harbor and discreetly reporting to Admiral Yamamoto. Japanese agents preferred to work

49. Kotani, Ken, "Japanese intelligence in WWII: Successes and Failures," accessed August 19, 2024, https://www.nids.mod.go.jp/english/publication/kiyo/pdf/2009/bulletin_e2009_2.pdf.

unseen, blending into local environments to gather sensitive information without drawing attention.

In contrast, American intelligence efforts in the Pacific Theater, led by the Office of Strategic Services (OSS) and the Office of Naval Intelligence (ONI), often involved larger, structured teams deployed in battalions or units. OSS operatives and Marine Intelligence officers conducted guerrilla warfare, sabotage, and reconnaissance across Pacific islands and mainland Asia, frequently coordinating in groups to carry out covert operations.[50] British intelligence agents in Shanghai, meanwhile, focused heavily on intercepting transmissions and monitoring local populations for unrest. Stationed in areas like the International Settlement, British operatives were responsible for intercepting enemy communications and using informants to counteract threats to British-controlled zones. Their intelligence gathering followed a more structured, European-style approach, with emphasis on maintaining stability within British territories by preemptively identifying potential security risks.

These varied operational styles allowed each agency to maximize its effectiveness, adapting approaches to meet the challenges of the Asia-Pacific Theater and ensure strategic intelligence gathering across multiple fronts.

---

50. Mark Harmon and Leon Carroll, *Ghosts of Honolulu*.

Chapter 3

Civilian Intelligence

Private Spies for the Wealthy and Influential

In addition to the extensive military intelligence that operated in Shanghai, civilian intelligence was also significant. Wealthy and influential families such as the Sassoon and Kadoorie families, both of Jewish origin and from Baghdad fled to Shanghai when Baghdad was still under the control of the Ottoman Empire and prior to the establishment of the modern Kingdom of Iraq, which was followed by the establishment of the Republic of Iraq in 1958. The presence of Jewish communities in Mesopotamia dates back to the Babylonian Exile around 598 BCE, when a portion of the Jewish population was taken to Babylon. After the Persian Emperor Cyrus the Great conquered Babylon in 539 BCE, he allowed these exiled Jews to return to Judah, though many chose to remain in Babylon. This community eventually became the thriving Baghdadi and Mashadi Jewish communities of modern-day Iraq and Iran, known for their involvement in regional trade routes like the Silk Road. The two families, having escaped persecution in Baghdad and settled in Shanghai, became quite influential, employing their own private intelligence agents which proved valuable in deciding the outcome of the war in the Pacific Theater in addition to playing a role in the European Theater. The two families had lived in Shanghai since the mid-19th century because they fled Baghdad due to Antisemitism and persecution by their non-Jewish neighbors. The Sassoon family originally fled to Bombay (modern day Mumbai) in India before moving to Shanghai, eventually becoming involved in the opium trade. The Kadoorie family was one of the primary families involved in the rubber trade as well as the early automobile trade in China. After the end of the Opium Wars, the two families had become very powerful in Shanghai.[51] Despite not speaking English or any of the various Chinese dialects, the Baghdadi-Jewish families operated two of the most famous hotels on the Bund in the International Zone: the Palace Hotel and the Cathay Hotel (now combined into the modern day Peace Hotel on the Bund)—which hosted important figures like Nationalist leader Chiang Kai-Shek.[52] As a result of coming into contact with so many highly influential figures, the two families employed private civilian spies in addition to their servants, hotel workers, and other employees to uncover information about their guests, allies, and enemies.[53] While they did not employ formal intelligence agents, their high-profile connections

51. Kaufman, Jonathan. *The Last Kings of Shanghai: The Rival Jewish Dynasties That Helped Create Modern China*. London: Penguin Books, 2021.
52. Ibid.
53. Ibid.

and visibility enabled them to navigate complex social and political environments, indirectly impacting wartime dynamics. The Sassoons and Kadoories left lasting legacies in Shanghai, with landmarks such as the Peace Hotel serving as enduring symbols of their influence.

The Righteous Among the Nations

While Italy was a member of the Axis Powers, they operated differently from their counterparts such as Germans and the Japanese and as a result, Italy was one of the easier places of occupied Europe to escape from, though it was still quite difficult.[54] With an influx of refugees from Europe, the Hongkou district was also called the "Restricted Sector for Stateless Refugees" and "Little Vienna" due to the populations that settled there with the help of the Sassoons, Kadoories, and Chinese diplomats such as Ho Feng Shan, known also by the title "The Chinese Schindler" due to his efforts to provide visas to Viennese Jews escaping Austria.[55] With connections to those in power, the Sassoon and Kadoorie families were able to obtain access to information, intentionally or otherwise, that much of the population could not. Between their status and network of civilian operatives from all walks of life, they were able have eyes and ears all throughout the Inter-

---

54. Ibid.
55. "Feng-Shan Ho," *Righteous Among the Nations*, Yad Vashem, accessed July 25, 2024, https://www.yadvashem.org/righteous/stories/ho.html.

national Zone which proved immensely useful during and after the Battle of Shanghai to keep the civilians of Shanghai and the International Zone safe despite the intense fighting, destruction of important infrastructure, use of chemical and biological weapons of war, and high death counts.[56] Shanghai became one of the few places in the world where Jewish refugees escaping Nazi persecution could enter without a visa, making it a vital refuge for those fleeing Europe during the 1930s and 1940s.

For his work helping to relocate fleeing European Jews to Shanghai by providing otherwise difficult to get visa papers, Ho Feng Shan later received the title "Righteous Among the Nations". This title is reserved for non-Jews who took great steps to protect Jews during the Holocaust while much of the rest of the world was indifferent or hostile to the Jewish people. To receive the title "Righteous Among the Nations" is considered to be a great honor.[57] Much of the work to rescue Jews by the "Righteous Among the Nations" was assisted by civilian spies working covertly to help the Jews escape to places of refuge such as Shanghai. Despite the war-torn state of Shanghai and after the Battle of Shanghai, it was still considered a much safer place by many in comparison to staying in Europe. Escaping to the West proved

---

56. "15 May 1947 WAR (Chemical Corps), to: CINCFE (G-2) (Chief Chemical Officer for Fell) NR: W 98097," Pacific Asia War Archive, accessed August 17, 2024, https://pacificatrocitiesedu.reclaim.hosting.
57. "Who Are the Righteous among the Nations?," Yad Vashem, accessed August 17, 2024, https://www.yadvashem.org/righteous/about-the-righteous/who-are-the-righteous-among-the-nations.html.

difficult, if not impossible, due to quotas instituted by the United States and Canada. Shanghai was one of the last places where extensive paperwork was not required, making it a desirable place of refuge despite the Japanese occupation and extensive destruction that was left in the wake of the Battle of Shanghai. Many refugees began to arrive in Shanghai in 1938 only months after the end of the Battle of Shanghai when the city was still much in ruins.[58] With the help of the Sassoons, the Kadoories, diplomats like Ho Feng Shan, and others, these refugees rebuilt their lives in Shanghai and came to help protect their new home from the onslaught of the Imperial Japanese Army as they descended to the International Zone in 1941. Since many Jewish refugees knew multiple languages, they were oftentimes recruited to assist with British or American intelligence operations, including code cracking and transmission interception between the Germans and the Japanese.[59] Additionally, some Japanese civilians of the International Zone did the same type of work but assisted the Japanese to intercept British, American, French, or Chinese telegram transmissions.

58. "Refugees from Pudong, arriving at the Bund by boat, Shanghai," 1937." *Historical Photographs of China*, accessed July 25, 2024, https://hpcbristol.net/visual/JM04-162.

59. "World War II: China—Shanghai." *The Jewish Community of China*, accessed June 19, 2024. http://www.jewsofchina.org/world-war-ii-china%E2%80%94shanghai.

Ho Feng Shan the "Chinese Schindler" and "Righteous Among the Nations".[60]

Civilians Skills for a Different Purpose

With almost nowhere else to turn, a significant amount of Jewish refugees opted to go east to Shanghai while others instead returned to their Indigenous homeland, Eretz Israel (The land of the People of Israel, לארשי ץרא)

60. "Ho Feng-Shan," *Jewish Virtual Library*, accessed September 8, 2024, https://www.jewishvirtuallibrary.org/feng-shan-ho.

in what was then called "The British Mandate of Palestine". This term had come into use after the Roman term "Syria-Palaestina" was coined to rename the Kingdom of Judea while attempting to sever Jewish ties to the land as punishment for the Bar Kokhba Revolt in 135 CE, the last of three revolts against the Romans. After the failed revolt and destruction of the second temple, the Jews were exiled from their homeland. Following their exile from Judea, many were enslaved by the Romans, forming a global diaspora scattered throughout Europe, Asia, and Africa. A significant population known as the Mizrahi Jews remained in the region and later helped those fleeing Europe to return to their indigenous homeland. After the establishment of the modern State of Israel in 1948 during the Expulsion of Jews from Arab Countries, the Farhud and other Antisemitic pogroms throughout Southwest Asia and North Africa, many Mizrahi who had lived as Dhimmis (non-Muslim second-class citizens) for centuries returned to their homeland alongside those who had never left and those who had more recently returned from the Ashkenazi and Sephardi diaspora. Many of the refugees fleeing Europe decided on joining those who had previously returned or otherwise never left and settled on Kibbutz (communal farms) in Jerusalem and throughout the Levant. Others who could not make the journey due to issues with papers or finding proper transport (such as trains or boats) fled to Shanghai. Shanghai offered a rare safe haven, accepting approximately 18,000 Jewish refugees.

A coin from the British Mandate period reading "Eretz Israel" in Hebrew in the left corner and Arabic in the right corner with the modern derivative of the Roman given term "Syria-Palestina" in English on top denoting the British Mandate oversight.[61]

While Jewish refugees in Shanghai primarily focused on survival, many of them supported their community in vital ways, often indirectly aiding wartime efforts. Wealthy families like the Sassoons and Kadoories, as well as Chinese diplomats such as Ho Feng Shan, provided essential support for the influx of Jewish refugees escaping Nazi persecution. These refugees found sanctuary in Shanghai's Hongkou district, where they developed robust mutual aid networks, including soup kitchens, schools, and community organizations that helped sustain thousands of displaced people during the Japanese occupation.

While Jewish refugees in Europe often contributed directly to intelligence operations, such as the Ritchie Boys in the European theater, there is limited evidence that Shanghai's Jewish refugees had formal roles in intelligence or collaborated directly with the OSS or other

61. "Ancient Jewish Coins," *Rebirth of Israeli Coins*, accessed September 8, 2024, https://www.jewishvirtuallibrary.org/rebirth-of-israeli-coins.

military intelligence in East Asia. However, informal community networks and established families like the Sassoons and Kadoories may have had access to useful local information through their business and social ties in the International Settlement.[62] In the Pacific Theater, the OSS's Detachment 404 was active in intelligence and guerrilla operations, particularly in the China-Burma-India Theater. Here, the OSS organized and trained local resistance forces, often collaborating with indigenous populations. While these civilian efforts were crucial, they were mostly separate from the refugee community's activities in Shanghai, which focused primarily on maintaining communal safety and resilience under challenging wartime conditions.[63]

While some civilian spies in Shanghai worked as intelligence agents to protect the city or aid Allied interests, others collaborated with Japanese forces. Civilian intelligence efforts in Shanghai were multi-faceted; some agents aligned with religious missions or private organizations with specific agendas, while others operated independently within the city's intricate social and political networks.

Religious missions and private citizens, including missionaries, sometimes provided intelligence infor-

62. Decker, Michael H., and Mackenzie, William, "The Birth and Early Years of Marine Intelligence," Marine Corps University, accessed August 17, 2024, https://www.usmcu.edu/Outreach/Marine-Corps-University-Press/MCH/Marine-Corps-History-Winter-2019/The-Birth-and-Early-Years-of-Marine-Corps-Intelligence/.

63. "Detachment 404," OSS Office of Strategic Services, accessed August 18, 2024, https://www.soc.mil/OSS/det-404.html.

mally, often through humanitarian or personal networks that gave them unique access to sensitive areas. U.S. Marine Corps Intelligence, in partnership with the Office of Naval Intelligence, established networks in Shanghai as early as the 1930s. Operatives like Major William Worton leveraged relationships with local informants and Chinese officials, who provided critical intelligence on Japanese movements. British intelligence similarly engaged operatives within Shanghai's International Settlement, gathering information to counter Japanese influence.

Meanwhile, Japanese intelligence, primarily through the Kempeitai and associated agents, conducted active surveillance in Shanghai. They infiltrated civilian and diplomatic circles, working to counter Allied intelligence efforts and securing information useful to Japan's military objectives.

This diverse web of affiliations highlights Shanghai's significance as a critical intelligence hub during the Asia-Pacific conflicts, with agents and operatives navigating complex allegiances and operating through various social and political channels.

Conclusion

The Fall of Shanghai and the End of the Asia-Pacific War

After the Battle of Shanghai and the defeat of the Chinese Army led by Chiang Kai-shek, the Imperial Japanese continued their expansion across China, imposing their rule on those who fell under their control, including both Han Chinese and various ethnic minorities. Sinicization—the process by which Han Chinese customs and language are introduced to integrate non-Han minority groups, such as the Manchu, Hui, or Miao—had long influenced these groups' identities. Aware of this, Japan attempted to capitalize on these ethnic distinctions to sway minority groups to their side during the Asia-Pacific War.

However, Japan's efforts met with limited success. Although the Japanese promoted preferential policies to appeal to groups like the Hui Muslims, emphasizing shared religious and cultural identities, the Hui largely maintained loyalty to the Chinese resistance. While some Manchu individuals collaborated with Japanese authorities in Manchukuo due to historical ties to the former Qing dynasty, this cooperation was not widespread among other minority groups, who resisted aligning with Japan.

The Japanese found few willing allies among China's ethnic minorities, particularly in regions with deep-rooted nationalist loyalties. Japan ultimately resorted to demonstrating military dominance to further its objectives, focusing on occupation rather than building substantial alliances. This aggressive approach led to notorious events like the Rape of Nanjing, even as Chinese forces regrouped to continue their resistance against the occupation

In Manchukuo, the Japanese continued their use of forced labor, conducted brutal experiments, and subjugated the local population through harsh conditions and servitude. Unit 731, a division of the Japanese Kwantung Army, notoriously conducted biological and chemical warfare experiments on Chinese civilians and prisoners of war in Harbin. These experiments, often resulting in death, included disease injections, hypothermia testing, and dissections. The atrocities committed in Manchukuo and elsewhere served as powerful propaganda for the Chinese resistance, fueling Chinese nationalism and uniting the people against Japanese occupation.

Although loyalty to the Empire of Japan was fervent on the Japanese home islands, morale and nationalism within Manchukuo were notably weaker. The puppet state's lack of genuine sovereignty, along with its exploitation by Japanese authorities, led to disillusionment among local officials and populations, who saw Japanese control as disenfranchising. This division bolstered the Chinese nationalist cause, as the population increasingly resisted Japanese rule, aided by covert Allied intelligence support.

The Battle of Shanghai was devastating for Chinese forces but contributed to a surge in Chinese nationalism. Despite the eventual fall of Shanghai, the hardships endured by the Chinese Army inspired a stronger sense of cultural pride and a commitment to push back against Japanese expansion throughout the region.

The Chinese Army continued to fight the Empire of Japan for eight more years until 1945 at which point the Empire of Japan surrendered to the Americans after the United States dropped atomic bombs on Hiroshima and Nagasaki. During this same period, the Soviet Red Army advanced to invade Manchuria with the collapse of Manchukuo after the surrender of Japan to the United States.[64] This surrender ended the Second World War, a war which resulted in approximately 15 to 20 million deaths in the European Theater with total death count estimates ranging from 35 to 60 million, including 6 million Jews being murdered in the Holocaust along with 6 million others. In the Pacific Theater, the death count included over 4 million Allied soldiers, 2.5 million soldiers from the Empire of Japan and its puppet states killed, and over 27 million civilian casualties (most of which occurred between 1937 and 1945). Tragically, Japan has never formally apologized in any notable capacity beyond a sur-

64. "Manchukuo." *Encyclopedia Britannica*, November 17, 2023. https://www.britannica.com/place/Manchukuo.

render ceremony to begin repatriation efforts nor have they expressed much repentance for their actions in colonizing Mainland China, Taiwan (Formosa), Korea, South Sakhalin, Thailand, Burma (Myanmar), Hong Kong, Vietnam, Singapore, Indonesia, Malaysia, Cambodia, Laos, The Philippines, East Timor, The Dutch East Indies, New Guinea, Guam, Wake Island, and Kiribati, or for attacking other islands without the intent to colonize, such as the Midway Atoll in the Pacific or the Allied base at Darwin harbor in northern Australia.[65]

The Loss of Manchukuo

Following the Battle of Shanghai and Japan's continued expansion across China, Japanese forces maintained a harsh occupation, particularly in Manchukuo. In Manchukuo, the Japanese regime intensified forced labor, conducted medical experiments, and exploited local resources for the war effort. By August 1945, as World War II neared its end, the Soviet Union declared war on Japan and launched a full-scale invasion of Manchukuo, overwhelming the Japanese Kwantung Army. This invasion hastened Japan's defeat, as many Japanese forces in Manchukuo surrendered or were repatriated following the war.

Japanese Emperor Puyi, a puppet ruler in Manchukuo, was captured by Soviet forces as he attempted

65. "The Demise of Imperial Japan," *Encyclopedia Britannica*, July 23, 2024, https://www.britannica.com/place/Empire-of-Japan/The-demise-of-imperial-Japan.

to escape to Japan. He was subsequently detained and charged with war crimes, spending a decade in a Chinese prison before his release. The Soviet presence in Manchukuo also helped diminish Japanese influence in the region, ultimately leading to the repatriation of many Japanese soldiers and civilians under Allied coordination efforts between 1946 and 1948.

After the dissolution of the puppet state of Manchukuo following Japan's defeat in World War II, Manchuria became a strategic base for the People's Liberation Army (PLA) under Mao Zedong in the resumed civil war against Chiang Kai-shek's National Revolutionary Army (NRA). While the Nationalists and Communists had been temporarily united against Japan during the Second Sino-Japanese War, their alliance dissolved after the Japanese surrender, reigniting hostilities. With Manchuria's industrial resources, strategic location, and Soviet support, the PLA was able to gain significant advantages, leading to critical victories over the Nationalists.

By late 1949, the NRA was defeated, and Chiang Kai-shek's government retreated to Taiwan—an island off mainland China known as Formosa from the Portuguese term meaning "beautiful island." Taiwan subsequently became the base for the Republic of China (ROC) under Chiang's leadership, though the People's Republic of China (PRC) was officially established on the mainland under Communist rule. These events marked the end of Nationalist control on the mainland and the establishment of Taiwan as the ROC, separate from the newly formed PRC on the mainland.[66] After Japan's defeat and the dissolution of its puppet state of Manchukuo, approximately 1.5 million Japanese civilians and soldiers were left stranded in China, primarily in Manchuria. The repatriation of these individuals was a complex process involving Chinese Nationalist forces, U.S. oversight, and logistical support at Huludao, a port in Liaoning Province. Between 1946 and 1948, about 1,050,000 Japanese were repatriated through Huludao in challenging conditions; many suffered or died from harsh winter weather and limited resources before reaching Japan, with some buried in nearby mountain sites. Some Japanese individuals were sent by Soviet forces to labor camps in Siberia, though there is limited evidence that Japanese soldiers directly assisted Chinese Communist forces against the Nationalists. Instead, most Japanese civilians and soldiers focused on surviving the harsh post-war environment and awaiting repatriation through U.S.-led efforts.[67]

The Escape of the Kempeitai

At the end of the war in Europe where some German and other collaborators were to be tried and punished for their crimes or to be recruited by foreign governments for top secret projects including the United States, many

66. Copper, John C., "Taiwan." *Encyclopedia Britannica*, August 16, 2024. https://www.britannica.com/place/Taiwan.
67. Maruyama, Paul K., *Escape from Manchuria: The Rescue of 1.7 Million Japanese Civilians Trapped in Soviet-Occupied Manchuria Following the End of World War II*. Kelly, NC: Toplink Publishing, 2017.

Nazi war criminals escaped to South America, especially Argentina. Some were hunted down at later dates either by British intelligence, the newly formed American Central Intelligence Agency, or the Israeli Mossad. Adolf Eichmann was captured in this way in Argentina by the Mossad during Operation Eichmann on May 11th, 1960.[68] While Eichmann and other escaped Nazi war criminals and ex-scientists were tracked down, captured, tried and executed for their crimes, this was not the case for many Japanese officials. A significant portion of the scientists and intelligence operatives who worked for the Imperial Japanese and evaded capture were not tried for their crimes or recruited by the Allied Forces.

Unlike the Germans and their collaborators, many Japanese military commanders and intelligence operatives successfully escaped China to Japan and were debriefed before being retired from service. The United States Office of Strategic Service, the precursor to the Central Intelligence Agency, along with The Office of Naval Intelligence continued their covert intelligence operations in Asia after the end of the Asia-Pacific War with the rise of the Cold War. British Intelligence continued to operate in Hong Kong due to its status as a British territory.[69]

---

68. "Operation Eichmann. Adolf Eichmann Was Captured in Argentina on May 11, 1960," Yad Vashem, accessed August 19, 2024, https://www.yadvashem.org/yv/en/exhibitions/eichmann/operation-eichmann.asp.
69. "The Office of Strategic Services: America's First Intelligence Agency," Central Intelligence Agency, accessed August 16, 2024, https://www.cia.gov/legacy/museum/exhibit/the-office-of-strategic-services-n-americas-first-intelligence-agency/.

The Importance of Intelligence

While there were many factors that impacted the Battle of Shanghai, the impact of intelligence is seen on all fronts: from civilian intelligence agents operating to protect or destroy the port city and its interests, foreign governments collecting valuable information on allied and enemy forces that turned the tide of the war, and Chinese and Japanese operatives searching out tactical advantages to exploit in the international city. While many documents were not declassified for decades, having access to these documents now helps historians to piece together what went on during the Battle of Shanghai from multiple perspectives. This historical information coming from primary sources gives a clearer picture as to what life was like for civilians, both those living in the city and those living in the International Settlement, plus intelligence agents operating in the shadows of the massive city, and combatants fighting to defend or destroy Shanghai, in what would be eventually known as one of the longest continuous battles in modern history.

Works Cited

"15 May 1947 WAR (Chemical Corps), to: CINCFE (G-2) (Chief Chemical Officer for Fell) NR: W 98097," Pacific Asia War Archive, accessed August 17, 2024, https://pacificatrocitiesedu.reclaim.hosting.

"1942: February 19: Japanese Raid at Darwin, Australia," accessed August 18, 2024, https://www.history.navy.mil/content/history/museums/nmusn/explore/photography/wwii/wwii-pacific/us-entry-into-wwii-japanese-offensive/1942-January-15-abda-japanese-octopus/1942-february-19-japanese-raid-darwin-australia.html.

"Between World Wars," *American Military History*, Chapter 19, accessed August 17, 2024, https://www.history.army.mil/books/AMH/amh-19.htm.

"Central China Area Operations Record, 1937-1941", Office of the Chief of Military History Department of the Army, accessed July 4, 2024, http://ibiblio.org/hyperwar/Japan/Monos/pdfs/JM-179/JM-179.pdf.

"Citizens of Shanghai Fleeing into Shanghai International Settlement—Battle of Shanghai (1937)," PICRYL—Public Domain Media Search Engine, https://jenikirbyhistory.getarchive.net/media/citizens-of-shanghai-fleeing-into-shanghai-international-settlement-battle-6b4b10.

Clark, Tim, "The Bushido Code: The Eight Virtues of the Samurai," accessed August 16, 2024, https://china.usc.edu/sites/default/files/forums/Samurai%20and%20the%20Bushido%20Code.pdf.

Copper, John C., "Taiwan." *Encyclopedia Britannica*, August 16, 2024. https://www.britannica.com/place/Taiwan.

"Counter-espionage summary—Shanghai," Central Intelligence Agency, accessed August 19, 2024, https://www.cia.gov/readingroom/docs/DOC_0000112821.pdf.

Decker, Michael H., and Mackenzie, William, "The Birth and Early Years of Marine Intelligence," Marine Corps University, accessed August 17, 2024, https://www.usmcu.edu/Outreach/Marine-Corps-University-Press/MCH/Marine-Corps-History-Winter-2019/The-Birth-and-Early-Years-of-Marine-Corps-Intelligence/.

"Detachment 202," OSS Office of Strategic Services, accessed August 18, 2024, https://www.soc.mil/OSS/det-202.html.

"Detachment 404," OSS Office of Strategic Services, accessed August 18, 2024, https://www.soc.mil/OSS/det-404.html.

"Establishment of Manchukuo," Pacific Atrocities Education, accessed August 12, 2024, https://www.pacificatrocities.org/establishing-manchukuo.html.

Fenby, Jonathan, *Chiang Kai Shek: China's Generalissimo and the Nation He Lost*. United States: Hachette Books, 2009.

"Feng-Shan Ho," *Righteous Among the Nations*, Yad Vashem, accessed July 25, 2024, https://www.yadvashem.org/righteous/stories/ho.html.

GHQ US Army Forces, Pacific Military Intelligence Section, "Saionji-Harada Memoirs Part 2: Mukden to Shanghai," *Pacific Asia War Archive*, accessed July 16, 2024, https://pacificatrocitiesedu.reclaim.hosting/admin/items/show/821.

Hall, Simon, "Imperial Japanese Army Intelligence in North and Central China During the Second Sino-Japanese War," *Salus Journal,* Vol. 2, Issue 2 (2014), pp. 16-30. Office of Justice Programs, accessed August 18, 2024, https://www.ojp.gov/ncjrs/virtual-library/abstracts/imperial-japanese-army-intelligence-north-and-central-china-during.

Harmon, Mark, and Carroll, Leon, *Ghosts of Honolulu: A Japanese Spy, a Japanese American Spy Hunter, and The Untold Story of Pearl Harbor*. New York: Harper Select, 2023.

Henriot, Christian, "August 1937: War and the Death En Masse of Civilians: Virtual Shanghai," *Virtual Shanghai*, accessed August 18, 2024, https://www.virtualshanghai.net/Texts/Articles?ID=130.

"Ho Feng-Shan," *Jewish Virtual Library*, accessed September 8, 2024, https://www.jewishvirtuallibrary.org/feng-shan-ho.

"Japan, China, the United States and the Road to Pearl Harbor, 1937-41," U.S. Department of State, accessed August 16, 2024, https://history.state.gov/milestones/1937-1945/pearl-harbor.

"'Japan Surrenders!': The National World War II Museum reminds Americans of the 65th anniversary of V-J Day and end of WWII," The National WWII Museum, New Orleans, August 10, 2010, https://www.nationalww2museum.org/media/press-releases/japan-surrenders-1.

Kaufman, Jonathan. *The Last Kings of Shanghai: The Rival Jewish Dynasties That Helped Create Modern China.* London: Penguin Books, 2021.

Kotani, Ken, "Japanese intelligence in WWII: Successes and Failures," accessed August 19, 2024, https://www.nids.mod.go.jp/english/publication/kiyo/pdf/2009/bulletin_e2009_2.pdf.

L'ultimo imperatore, Roma: Gruppo editoriale L'Espresso, 2003.

"Manchukuo." *Encyclopedia Britannica*, November 17, 2023. https://www.britannica.com/place/Manchukuo.

"Marco Polo Bridge Incident," *Encyclopedia Britannica*, June 30, 2024, https://www.britannica.com/event/Marco-Polo-Bridge-Incident.

Maruyama, Paul K., *Escape from Manchuria: The Rescue of 1.7 Million Japanese Civilians Trapped in Soviet-Occupied Manchuria Following the End of World War II*. Kelly, NC: Toplink Publishing, 2017.

Noble, Dennis L., "A US Naval Intelligence Mission to China in the 1930s," *Studies In Intelligence* Vol. 50 No. 2 (2006)," Central Intelligence Agency, accessed August 17, 2024, https://www.cia.gov/resources/csi/studies-in-intelligence/volume-50-no-2/a-us-naval-intelligence-mission-to-china-in-the-1930s/.

"Operation Eichmann. Adolf Eichmann Was Captured in Argentina on May 11, 1960," Yad Vashem, accessed August 19, 2024, https://www.yadvashem.org/yv/en/exhibitions/eichmann/operation-eichmann.asp.

Paulose, James. "Three Months of Bloodshed: Strategy and Combat During The Battle of Shanghai." *Report: West Point*

Undergraduate Historical Review 3, Vol. 2 (2013): 9-22, https://athena.westpoint.edu/items/ca1491c1-b912-49e7-9e38-8ec9867d13c4.

Pua, Derek, Dybbro, Danielle, and Rogers Alistair, *Unit 731: The Forgotten Auschwitz* San Francisco, CA: Pacific Atrocities Education, 2018.

Rawski, Evelyn S., "Reenvisioning the Qing: The Significance of the Qing Period in Chinese History," *Journal of Asian Studies* 55(4), doi:10.2307/2646525.

Reference Division, "Developments in the Economic Structure of the Co-Prosperity Sphere During the Past Year Part II—The Outer Zone," *Pacific Asia War Archive*, accessed June 12, 2024, https://pacificatrocitiesedu.reclaim.hosting/admin/items/show/714.

"Refugees from Pudong, arriving at the Bund by boat, Shanghai," 1937." *Historical Photographs of China*, accessed July 25, 2024, https://hpcbristol.net/visual/JM04-162.

Reynolds, Nicholas, *Need to Know: World War II and the Rise of American Intelligence*. Boston: Mariner Books, 2023.

"Secret Intelligence," OSS Office of Strategic Services, accessed August 18, 2024, https://www.soc.mil/OSS/secret-intelligence.html.

"SIS—Our History," accessed August 19, 2024, https://www.sis.gov.uk/our-history.html.

"The China Marines | History Detectives," PBS, accessed August 17, 2024, https://www.pbs.org/opb/historydetectives/feature/the-china-marines/.

"The Chinese Revolution of 1911," Office of the Historian, U.S. Department of State, accessed April 27, 2024, https://history.state.gov/milestones/1899-1913/chinese-rev.

"The Demise of Imperial Japan," *Encyclopedia Britannica*, July 23, 2024, https://www.britannica.com/place/Empire-of-Japan/The-demise-of-imperial-Japan.

"The First Battle of Shanghai 1932," Pacific Atrocities Education, accessed August 16, 2024, https://www.pacificatrocities.org/the-first-battle-of-shanghai-1932.html.

"The Office of Strategic Services: America's First Intelligence Agency," Central Intelligence Agency, accessed August 16, 2024, https://www.cia.gov/legacy/museum/exhibit/the-office-of-strategic-services-n-americas-first-intelligence-agency/.

"The United States Marines in North China," US Marine Corps, accessed August 23, 2024, https://www.usmcu.edu/Portals/218/TheUnitedStatesMarinesInNorthChina1945-1949.pdf.

"U.S. Armed Forces in China," *Dictionary of American History*, Encyclopedia.com, August 17, 2024, https://www.encyclopedia.com/history/dictionaries-thesauruses-pictures-and-press-releases/china-us-armed-forces.

"US Marines Intelligence Report 5 December, 1935," accessed July 4, 2024, https://www.pacificatrocitiesedu.reclaim.hosting/items/show/221.

"Victims by Bombing of Chinese Bombers, Battle of Shanghai, 1937," PICRYL—Public Domain Media Search Engine, accessed October 12, 2024.

https://jenikirbyhistory.getarchive.net/media/victims-by-bombing-of-chinese-bombers-battle-of-shanghai-1937-ce0c07.

"Victims of "Bloody Saturday" bombing, Ave Edward VII, Shanghai, 14 August 1937," *Historical Photographs of China*, accessed June 24, 2024, https://hpcbristol.net/visual/AL-s63.

"What Was the Battle of Verdun?," Imperial War Museums, accessed August 12, 2024, https://www.iwm.org.uk/history/what-was-the-battle-of-verdun.

"What Was the Holocaust?" Yad Vashem, accessed August 14, 2024, https://www.yadvashem.org/holocaust/about.html.

"Who Are the Righteous among the Nations?," Yad Vashem, accessed August 17, 2024, https://www.yadvashem.org/righteous/about-the-righteous/who-are-the-righteous-among-the-nations.html.

"World War II: China—Shanghai." *The Jewish Community of China*, accessed June 19, 2024. http://www.jewsofchina.org/world-war-ii-china%E2%80%94shanghai.

Yang, Liping, "Researching the History of Shanghai Between the 1830s and 1950s," *The Gale Review*, accessed June 19, 2024, https://review.gale.com/2023/12/05/researching-the-history-of-shanghai-between-the-1830s-and-1950s/.

"Yangtze River Patrol and Other US Navy Asiatic Fleet Activities in China, 1920-1942, as Described in the *Annual Reports of the Navy Department*," Naval History and Heritage Command, accessed August 17, 2024, https://www.history.navy.mil/research/library/online-reading-room/title-list-alphabetically/y/yangtze-river-patrol-and-other-us-navy-asiatic-fleet-activities-in-china.html.

Made in the USA
Monee, IL
04 November 2026

67386121R10046